The Elements, Proton-Seconds, And Quanta

Ian Beardsley
(*University of Oregon, Department of Physics, 2022*)

Genesis Project · California · 2022

Abstract: It would seem to not only are artificial intelligence elements mathematical constructs, but so are biological life elements, and they are described in terms of one another, this is the subject of part 1. It becomes necessary to say if these elements are subject to mathematical constructs then matter, or inertia is a mathematical construct. This results in a constant that not only predicts the hydrocarbons, the skeletons of life, but determines a definition for the radius of the solar system and predicts the radius of the proton. This is part 2. In part 3 we show this takes the form of a gaussian distribution meaning it can all be solved with wave mechanics. In part 4 I show my first development the lead to all of this. In Part 5, since all of this is related to the calendar as developed from ancients times leading to the duration of a second I look at eh archaeology of other star systems.

Part 1 The Mathematical Construct................4

Part 2 The Constant....................................10

Part 3 The Wave Packet.............................33

Part 4 Where It Began...............................42

Part 5 Archeology of other Star Systems.........50

Part 1: The Mathematical Construct

Building A Matrix

We pull these Al elements out of the periodic table of the elements to make an Al periodic table:

```
|  13   14   15
2  B
3  Al   Si   P
4  Ga   Ge   As
```

We now notice we can make a 3 by 3 matrix of it, which lends itself to to the curl of a vector field by including biological elements carbon C (above Si):

$$\begin{pmatrix} \vec{i} & \vec{j} & \vec{k} \\ \frac{\partial}{\partial x} & \frac{\partial}{\partial y} & \frac{\partial}{\partial z} \\ (-C \cdot P)y & (Si \cdot Ga)z & (Ge \cdot As)y \end{pmatrix} =$$

$$(Ge \cdot As - Si \cdot Ga)\vec{i} + (C \cdot P)\vec{k} =$$

$$[(72.64)(74.92) - (28.09)(69.72)]\,\vec{i} + [(12.01)(30.97)]\,\vec{k}$$

Which resulted in Stokes theorem (Beardsley, Essays In Cosmic Archaeology Volume 3):

Equation 1.
$$\sqrt[5]{\int_{Si}^{Ge}\int_{Si}^{Ge} \nabla \times \vec{u} \cdot d\vec{a}} = exp\left(\frac{1}{Ge - Si}\int_{Si}^{Ge} ln(x)dx\right)$$

Where

$$\nabla \times \vec{u} = (Ge \cdot As - Si \cdot Ga)\vec{i} + (C \cdot P)\vec{k}$$

$$d\vec{a} = \left(zdydz\vec{i} + ydydz\vec{k}\right)$$

$$\vec{u} = (-C \cdot P)y\vec{i} + (Si \cdot Ge)z\vec{j} + (Ga \cdot As)y\vec{k}$$

We were then able to write this with product notation

Equation 2.
$$\sqrt[5]{\int_{Si}^{Ge}\int_{Si}^{Ge} \nabla \times \vec{u} \cdot d\vec{a}} = \sqrt[n]{\prod_{i=1}^{n} x_i}$$

While we have the Al BioMatrix

B.   C.   N.

Al.   Si.   P.

Ga.   Ge.   As.

Which we used to formulate a similar equation (Beardsley, Essays In Cosmic Archaeology Volume 2)

We can form another 3X3 matrix we will call the electronics matrix (Beardsley, Cosmic Archaeology, Volume Three):

Ni.   Cu.   Zn.

Pd.   Ag.   Cd.

Pt.   Au.   Hg.

We can remove the 5th root sign in the above equation by noticing

Equation 3.
$$\prod_{i=1}^{5} x_i = Si \cdot Ge \cdot Cu \cdot Ag \cdot Au$$

=(28.085)(72.64)(12.085)(107.8682)(196.9657)=

$$523{,}818{,}646.5 \frac{g^5}{mol^5}$$

Where we have substituted carbon (C=12.01) the core biological element for copper (Cu).

But since we have:

Equation 4. $$\int_{Si}^{Ge}\int_{Si}^{Ge}(\nabla\times\vec{u})\cdot d\vec{a} = 170,535,359.662 (g/mol)^5$$

We take the ratio and have

$$\frac{523,818,646.5}{170,535,359.662} = 3.0716$$

Almost exactly 3 which is the ratio of the perimeter of regular hexagon to its diameter used to estimate pi in ancient times by inscribing it in a circle:

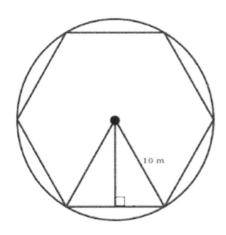

Perimeter=6

Diameter=2

6/2=3

$\pi = 3.141...$

Thus we have the following equation…

Equation 5. $$\pi\int_{Si}^{Ge}\int_{Si}^{Ge}(\nabla\times\vec{u})\cdot d\vec{a} = \prod_{i=1}^{5} x_i$$

Showing The Calculation using the most accurate data possible…

Ge=72.64
As=74.9216
Si=28.085
Ga=69.723
C=12.011
P=30.97376200

$(Ge \cdot As - Si \cdot Ga)\vec{i} + (C \cdot P)\vec{k} =$

$[(72.64)(74.9216) - (28.085)(69.723)]\vec{i} + [(12.011)(30.97376200)]\vec{k} =$

$3,484.134569\left(\dfrac{g}{mol}\right)^2 \vec{i} + 372.025855\left(\dfrac{g}{mol}\right)^2 \vec{k}$

$\displaystyle\int_{Si}^{Ge}\int_{Si}^{Ge}\left(3,484.134569\left(\dfrac{g}{mol}\right)^2 \vec{i} + 372.025855\left(\dfrac{g}{mol}\right)^2 \vec{k}\right) \cdot \left(zdydz\vec{i} + ydydz\vec{k}\right)$

$\displaystyle\int_{Si}^{Ge}\int_{Si}^{Ge}\left(3,484.134569\left(\dfrac{g}{mol}\right)^2 \cdot zdzdy + 372.025855\left(\dfrac{g}{mol}\right)^2 \cdot ydzdy\right)$

$\displaystyle\int_{Si}^{Ge} 3,484.134569\left(\dfrac{(72.64 - 28.085)^2}{2}\right)dy + \int_{Si}^{Ge} 372.025855 y \cdot (72.64 - 28.085)dy$

$3458261.42924\left(\dfrac{g}{mol}\right)^4 (72.64 - 28.085) + 16575.6119695\left(\dfrac{g}{mol}\right)^3 \left(\dfrac{(72.64 - 28.085)^2}{2}\right)$

$= 154,082,837.980 + 16,452,521.6822 =$

$170,535,359.662\left(\dfrac{g}{mol}\right)^5$

$\displaystyle\prod_{i=1}^{5} x_i = Si \cdot Ge \cdot C \cdot Ag \cdot Au =$

$(28.085)(72.64)(12.085)(107.8682)(196.9657) =$

$523,818,646.5 \dfrac{g^5}{mol^5}$

Where we have substituted carbon C=12.01 for copper Cu. We use Cu, Ag, Au because they are the middle column of our electronics matrix, they are the finest conductors used for electrical wire. We use C, Si, Ge because they are the middle column of our AI Biomatrix. Si and Ge are the primary semiconductor elements used in transistor technology (Artificial Intelligence) and C is the core element of biological life. We have

$\dfrac{523,818,646.5}{170,535,359.662} = 3.0716$

$\pi = 3.141...$

Perimeter/Diameter of regular hexagon = 3.00

$$\frac{3.141 + 3.00}{2} = 3.0705$$

The same value as our 3.0716 if taken at two places after the decimal.

Part 2: The Constant

Introduction

In order to present the elements as mathematical structures we need to explain the matter from which they are made as mathematical constructs. We need a theory for Inertia. I had found (Beardsley Essays In Cosmic Archaeology. 2021) where I suggested the idea of proton seconds, that is six proton-seconds, which is carbon the core element of biological life if we can figure out a reason to divide out the seconds. I found

Equation 1. $$\frac{1}{t_1 \alpha^2 m_p} \sqrt{\frac{h 4\pi r_p^2}{Gc}} = 6 protons$$

Where h is Planck's constant 6.62607E−34 Js, r_p is the radius of a proton 0.833E-15m, G is the universal constant of gravitation 6.67408E-11 (Nm2)/(kg2), and c is the speed of light 299,792,459 m/s. And t_1 is t=1 second. α is the Sommerfeld constant (or fine structure constant) is 1/137. The mass of a proton is $m_p = 1.67262E - 27 kg$.

The fine structure constant squared is the ratio of the potential energy of an electron in the first circular orbit to the energy given by the mass of an electron in the Bohr model times the speed of light squared:

$$\alpha^2 = \frac{U_e}{m_e c^2}$$

Matter is that which has inertia. This means it resists change in position with a force applied to it. The more of it, the more it resists a force. We understand this from experience, but what is matter that it has inertia?

I would like to answer this by considering matter in one of its simplest manifestations, the proton, a small sphere with a mass of 1.6726E-27 kg. This is a measure of its inertia.

I would like to suggest that matter, often a collection of these protons, is the three dimensional cross-section of a four dimensional hypersphere.

The way to visualize this is to take space as a two-dimensional plane and the proton as a two dimensional cross-section of a sphere, which would be a circle.

In this analogy we are suggesting a proton is a three dimensional bubble embedded in a two dimensional plane. As such there has to be a normal vector holding the higher dimensional sphere in a lower dimensional space. Thus if we apply a force to to the cross-section of the sphere in the plane there should be a force countering it proportional to the normal holding it in a lower dimensional universe. This counter force would be experienced as inertia. It may even induce in it an electric field, and we can see how it may do the same equal but opposite for the electron. Refer to the illustration on the following page…

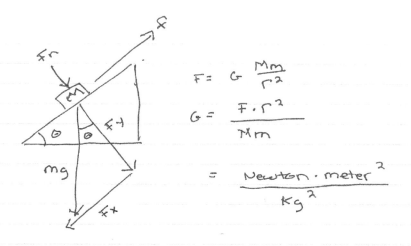

$$F = G\frac{Mm}{r^2}$$

$$G = \frac{F \cdot r^2}{Mm}$$

$$= \frac{Newton \cdot meter^2}{Kg^2}$$

G measures the amount of force produced per kilogram of mass over a distance in meters.

$F_f = \mu F_y \quad F_y = F\cos\theta \quad F_x = F\sin\theta$

the force f due to friction equals the normal force times the coefficient of friction μ.

$F_x = F\sin\theta - f \quad ma = mg\sin\theta - \mu mg\cos\theta$

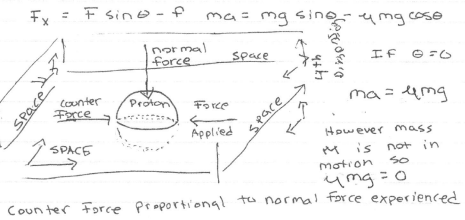

IF $\theta = 0$

$ma = \mu mg$

However mass M is not in motion so $\mu mg = 0$

Counter force proportional to normal force experienced as inertia. Matter as a three-dimensional cross-section of a four-dimensional hypersphere. It is 4-dimension space (bubble) crossection by 3D space

$$\frac{1}{\alpha^2 m_p}\sqrt{\frac{h 4\pi r_p^2}{Gc}} = (6 protons)(1 second)$$

$$\frac{1}{\alpha^2 m_p}\sqrt{\frac{h 4\pi r_p^2}{Gc}} = (1 proton)(6 seconds)$$

$$\alpha^2 = \frac{U_e}{m_e c^2}$$

I made a program that looks for close to whole number solutions, I set it at decimal part equal to 0.25. You can choose how may values for t you want to try, and by what to increment them. Here are the results for incrementing by 0.25 seconds then 0.05 seconds. Constant to all of this is hydrogen and carbon. The smaller integer value of seconds gives carbon (6 protons at 1 second) and the largest integer value of seconds gives hydrogen (1 proton at six seconds) and outside of that for the other integer values of protons you get are at t>0 and t<1. Equation 1 really has some interesting properties. Here are two runs of the program(decpart is just me verifying that my boolean test was working right to sort out whole number solutions):

By what value would you like to increment?: 0.25
How many values would you like to calculate for t in equation 1 (no more than 100?): 100
24.1199 protons 0.250000 seconds 0.119904 decpart
12.0600 protons 0.500000 seconds 0.059952 decpart
8.0400 protons 0.750000 seconds 0.039968 decpart
6.0300 protons 1.000000 seconds 0.029976 decpart
4.0200 protons 1.500000 seconds 0.019984 decpart
3.0150 protons 2.000000 seconds 0.014988 decpart
2.1927 protons 2.750000 seconds 0.192718 decpart
2.0100 protons 3.000000 seconds 0.009992 decpart
1.2060 protons 5.000000 seconds 0.205995 decpart
1.1486 protons 5.250000 seconds 0.148567 decpart
1.0964 protons 5.500000 seconds 0.096359 decpart
1.0487 protons 5.750000 seconds 0.048691 decpart
1.0050 protons 6.000000 seconds 0.004996 decpart
0.2487 protons 24.250000 seconds 0.248659 decpart
0.2461 protons 24.500000 seconds 0.246121 decpart
0.2436 protons 24.750000 seconds 0.243635 decpart

By what value would you like to increment?: 0.05
How many values would you like to calculate for t in equation 1 (no more than 100?): 100
40.1998 protons 0.150000 seconds 0.199837 decpart
30.1499 protons 0.200000 seconds 0.149879 decpart
24.1199 protons 0.250000 seconds 0.119904 decpart
20.0999 protons 0.300000 seconds 0.099918 decpart
17.2285 protons 0.350000 seconds 0.228500 decpart
15.0749 protons 0.400000 seconds 0.074938 decpart
12.0599 protons 0.500000 seconds 0.059950 decpart
10.0500 protons 0.600000 seconds 0.049958 decpart
8.0400 protons 0.750000 seconds 0.039966 decpart

7.0941 protons 0.850000 seconds 0.094088 decpart
6.0300 protons 1.000000 seconds 0.029975 decpart
5.2435 protons 1.150000 seconds 0.243457 decpart
5.0250 protons 1.200000 seconds 0.024980 decpart
4.1586 protons 1.450000 seconds 0.158605 decpart
4.0200 protons 1.500000 seconds 0.019985 decpart
3.1737 protons 1.899999 seconds 0.173673 decpart
3.0923 protons 1.949999 seconds 0.092296 decpart
3.0150 protons 1.999999 seconds 0.014989 decpart
2.2333 protons 2.699999 seconds 0.233325 decpart
2.1927 protons 2.749999 seconds 0.192719 decpart
2.1536 protons 2.799999 seconds 0.153564 decpart
2.1158 protons 2.849998 seconds 0.115782 decpart
2.0793 protons 2.899998 seconds 0.079303 decpart
2.0441 protons 2.949998 seconds 0.044061 decpart
2.0100 protons 2.999998 seconds 0.009993 decpart
1.2433 protons 4.850000 seconds 0.243294 decpart
1.2306 protons 4.900001 seconds 0.230607 decpart
1.2182 protons 4.950001 seconds 0.218177 decpart

Here is the code for the program:

```c
#include <stdio.h>
#include <math.h>
int main(int argc, const char * argv[]) {
    
    int n;
    float value=0, increment,t=0, p=1.67262E-27, h=6.62607E-34,G=6.67408E-11, c=299792459,protons[100],r=0.833E-15;
    
    do
    {
        printf("By what value would you like to increment?: ");
        scanf("%f", &increment);
    printf("How many values would you like to calculate for t in equation 1 (no more than 100?): ");
        scanf("%i", &n);
    }
        while (n>=101);
    {
        
        for (int i=0; i<n;i++)
        {
            protons[i]=((137*137)/(t*p))*sqrt(h*4*(3.14159)*(r*r)/(G*c));
            
            int intpart=(int)protons[i];
            float decpart=protons[i]-intpart;
            t=t+increment;
            if (decpart<0.25)
            { printf("%.4f protons %f seconds %f decpart \n", protons[i], t-increment, decpart);
            }}}}
```

While I understand that one second is a human invention and can't be taken as significant, the equations have meaning in that there are two equations each utilizing a second so they are connected. But what really makes me wonder is how one can predict carbon, and the other hydrogen so accurately with the unit of a second. It was a conspiracy on the part of those who formulated the duration of a second a long time ago to be what it is? The Ancient Egyptians, The Babylonians, The Julian calendar—who, what, when?

6 protons gives a little more than a second. This makes a shorter day. We have

Equation 2. $$\frac{1}{6\alpha^2 m_p}\sqrt{\frac{h4\pi r_p^2}{Gc}} = 1.004996352 seconds$$

H=1.00784 g/mol, carbon = 6 protons
h=6.62607E-34, r_p=0.833E-15, G=6.67408E-11, c=299,792,459

The Second

But actually the second might have physical meaning beyond what is here. Not only is the second related to the earth orbital period but, to other things; By dividing the day into 24 hours, the hour into 60 minutes, and the minute into 60 seconds, the second is 1/86400 of day. By doing this we have a twelve-hour daytime at spring and fall equinox on the equator, 12 being the most divisible number for its size (smallest abundant number). That is to say that twelve is evenly divisible by 1,2,3,4,6 which precede it and 1+2+3+4+6=16 is greater than twelve. As such there is about one moon per 30 days and about 12 moons per year (per each orbit) giving us a twelve-month calendar. This is all further convenient in that the moon and earth are in very close to circular orbits and the circle is evenly divisible by 30, 45, 60, and 120 if we divide the circle into 360 degrees which are special angles very useful to the workings of physics and geometry. Further, the 360 degrees of a circle are about the 365 days of a year (period of one earth orbit) so as such the earth moves through about a degree a day in its journey around the sun. Thus, through these observations down through the ages since ancient times we have constructed the duration of a second wisely enough to make a lot work together. Now we see 6 protons, which is carbon the core element of biological life on the planet where all of this came together is deeply connected with the second that defines it all. With this idea of proton-seconds describing hydrogen and carbon the basis of life the hydrocarbons in a cycle of 6 with respect to one another, the motion of the earth around the sun and moon around the earth, and the basis of geometry the 360 degree circle, equation 1 connects them with the universal constant of gravitation, the speed of light, the fine structure constant and Planck's constant that characterize the physics of the atom. I really wonder if other star systems are connected so well to their planets with their star as is the Earth with the Sun. We really can't resolve planets around other stars because the stars are too far and so bright compared to their orbiting planets and the planets have to be large enough to be inferred by the motion they induce in the star they orbit. It is hard to do so with earth sized planets that might harbor life.

The Geometrical Explanation of Seconds

Here we will talk about the equation 1:

$$\frac{1}{t_1 \alpha^2 m_p}\sqrt{\frac{h4\pi r_p^2}{Gc}} = 6 protons$$

In so far as the second unifies carbon (6 protons) with hydrogen (1 proton) through the unit of a second as the hydrocarbons the backbones of life. We have suggested the second is important as well in terms of the phases of the moon and the earth and that these determined the calendar system we use. We further suggested there is a connection of this to the structure found in geometry, and this is what we want to explore further, here. We ended with all of this can be compactly written as:

$$2cos(\pi/n) = 1, \sqrt{2}, \Phi, \sqrt{3}, \ldots$$

Where n=3, 4, 5, ,6 ,…

We could evaluate this for n equal to other integers, or even the numbers, but these produce the special triangles, and geometries we are most interested. Thus we will begin by pointing out that

3 X 4 X 5 X 6 = 360

The amount of degrees into which we divide a circle and that, as such it approximates the number of days in a year (1 revolution of the earth around the sun) and thus we see that

Equation 3. $\quad 2cos(\pi/n) = 1, \sqrt{2}, \Phi, \sqrt{3}, \ldots$

Represents days as well (The earth moves through about 1 degree a day in its orbit around the sun) by solving it for n:

Equation 4. $\quad days = cos^{-1}(y/2)$

Where,

$$y = 1, \sqrt{2}, \frac{\sqrt{5}+1}{2}, \sqrt{3}, \ldots$$

Which correspond respectively to:

$n = 3,4,5,6,\ldots$

Which are the unit triangle, unit the square, the unit regular pentagon, and the unit regular hexagon.

The Derivative

We take our function

$$days = cos^{-1}(y/2)$$

And write it:

$$y = cos^{-1}(x/2)$$

Then take the derivative:

Equation 5. $$\frac{dy}{dx} = -\frac{1}{\sqrt{4-x^2}}$$

This is a right triangle with hypotenuse 2, and height x and with base $\sqrt{4-x^2}$:

[handwritten diagram: right triangle with hypotenuse 2, height x, base $\sqrt{4-x^2}$, angle θ]

$\frac{\sqrt{4-x^2}}{2} = \cos\theta$

$\sqrt{4-x^2} = 2\cos\theta$

$\cos^{-1}\frac{\sqrt{4-x^2}}{2} = \theta$

To get $\theta = 60°$, then $x = \sqrt{3}$. To get $\theta = 45°$, then $x = \sqrt{2}$, and to get $\theta = 30°$ then $x = 1$.

To get $\theta = 36°$ then $\frac{\sqrt{4-x^2}}{2} = \Phi$. Thus our function is equation four written:

$y = 2\cos(\theta(n))$

Where:

$$\theta(n) = \frac{\pi}{n} = \frac{180°}{n}$$

n=3,4,5,6,,...

$$\frac{dy}{dx} = -\frac{1}{\sqrt{4-x^2}} =$$

$-1/\sqrt{3}$ is the rate of change of a triangle (x=1). -0.85 is the rate of change of a regular pentagon ($f : x \mapsto \Phi$). $-\frac{\sqrt{2}}{2}$ is the rate of change of a square ($x = \sqrt{2}$). -1 is the rate of change of a regular hexagon ($x = \sqrt{3}$).

We see that

$y = \cos^{-1}(x/2)$

$$\frac{dy}{dx} = -\frac{1}{\sqrt{4-x^2}}$$

Is x as a function of n, and that n=3 is a unit triangle, n=4 a unit square, n=5 is a unit pentagon, and n=6 is a unit hexagon. Thus if n=3 we have the unit triangle is Earth, the unit square is

Mars, the unit pentagon is n=5 which not only is this shape not a member of the regular tessellators, n=5 is the asteroid belt, which is a location in the solar system where a planet cannot form. We then proceed to Jupiter, which is n=6, the most massive planet in the solar system which carries the majority of its mass. dy/dx is the the change in days with respect to planetary number.

The Formulation

We can actually formulate this differently than we have. We had

$$\frac{1}{t_1}\frac{1}{\alpha^2 m_p}\sqrt{\frac{h 4\pi r_p^2}{Gc}} = 6 protons$$

$$\frac{1}{t_6}\frac{1}{\alpha^2 m_p}\sqrt{\frac{h 4\pi r_p^2}{Gc}} = 1 proton$$

But if t1 is not necessarily 1 second, and t6 is not necessarily six seconds, but rather t1 and t2 are lower and upper limits in an integral, then we have:

Equation 6. $$\frac{1}{\alpha^2 m_p}\sqrt{\frac{h 4\pi r_p^2}{Gc}} \int_{t_1}^{t_2} \frac{1}{t^2} dt = \mathbb{N}$$

And we can do this:

$$6\int_{\sqrt{2}}^{\sqrt{3}} \frac{1}{t^2} dt = 6\left(\frac{1}{\sqrt{2}} - \frac{1}{\sqrt{3}}\right) = 0.7806$$

If we take our equation 4 and integrate over the same range:

$$\int_{\sqrt{2}}^{\sqrt{3}} \cos^{-1}(x/2)dx = \frac{1}{6}\left(\sqrt{3}\pi - 6\right) - \frac{\pi - 4}{2\sqrt{2}} = 0.21039$$

If we add these two the result we get is:

0.7806+0.21039=0.99~1.00

The Solar System And Sand

While we have

$$\frac{1}{\alpha^2 m_p}\sqrt{\frac{h4\pi r_p^2}{Gc}}\int_{t_1}^{t_2}\frac{1}{t^2}dt = \mathbb{N}$$

Is a number of protons

$$\frac{1}{\alpha^2 m_p}\sqrt{\frac{h4\pi r_p^2}{Gc}}$$

Is proton-seconds. Divide by time we have a number of protons because it is a mass divided by the mass of a proton. But these masses can be considered to cancel and leave pure number. We have that

$$6\int_{\sqrt{2}}^{\sqrt{3}}\frac{1}{t^2}dt = 6\left(\frac{1}{\sqrt{2}} - \frac{1}{\sqrt{3}}\right) = 0.78$$

$$\int_{\sqrt{2}}^{\sqrt{3}}\cos^{-1}(x/2)dx = \frac{1}{6}\left(\sqrt{3}\pi - 6\right) - \frac{\pi - 4}{2\sqrt{2}} = 0.21$$

Interestingly 78% is the percent of N2 in the atmosphere and 21% is the percent of O2 in the atmosphere (by volume). These are the primary constituents that make it up. The rest is primarily argon and CO2. This gives the molar mass of air as a mixture is:

$0.78N2 + 0.21O2 = 29.0 g/mol$

Now interestingly, I have found this connected to the solar system by considering a mixture of silicon, the primary constituent of the Earth crust, and germanium just below it in the periodic table, in the same proportions of 78% and 21%. Silicon is also the primary second generation semiconductor material (what we use today) and germanium is the primary first generation semiconductor material (what we used first). The silicon is directly below our carbon of one proton-second, silicon directly below that, and germanium directly below that, in the periodic table. So we are moving directly down the periodic table in group 14. The density of silicon is 2.33 g/cm3 and that of germanium is 5.323 g/cm3. Let us weight these densities with our 0.21 and 0.78:

$0.21Si + 0.78Ge = 4.64124 g/cm^3$

Now consider this the starting point for density of a thin disc decreasing linearly from the Sun to Pluto (49.5AU=7.4E14cm). Thus,…

$$\rho(r) = \rho_0\left(1 - \frac{r}{R}\right)$$

Thus,...

$$M = \int_0^{2\pi} \int_0^R \rho_0 \left(1 - \frac{r}{R}\right) r \, dr \, d\theta$$

Or,...

$$M = \frac{\pi \rho_0 R^2}{3}$$

Thus,...

$$M = \frac{\pi (4.64124)(7.4E14)^2}{3} = 2.661E30$$

If we add up the masses of the planets in our solar system they are 2.668E30 grams.

Since

$$\frac{2.661}{2.668}(100) = 99.736$$

Meaning mixing germanium and silicon in the same proportion that occurs with N2 and O2 in the atmosphere and with

$$6 \int_{\sqrt{2}}^{\sqrt{3}} \frac{1}{t^2} dt = 6 \left(\frac{1}{\sqrt{2}} - \frac{1}{\sqrt{3}}\right) = 0.78$$

$$\int_{\sqrt{2}}^{\sqrt{3}} \cos^{-1}(x/2) dx = \frac{1}{6}\left(\sqrt{3}\pi - 6\right) - \frac{\pi - 4}{2\sqrt{2}} = 0.21$$

Where

$$6 = \frac{1}{\alpha^2 m_p} \sqrt{\frac{h 4\pi r_p^2}{Gc}}$$

In the first integral. See the following pages to see the computation of the mass of the planets in the solar system...

Planet	In Earth Mass
Mercury.	0.0553
Venus.	0.815
Earth.	1
Mars.	0.11
Jupiter.	317.8
Saturn.	95.2
Uranus.	14.6
Neptune.	17.2
Moon.	0.0123

Earth Mass In Grams: 5.972×10^{27}

Asteroid Belt: 4% of Moon (0.00492)

AU: 1.496×10^{13} cm

Pluto (Edge of Solar System): 49.5 AU (7.4×10^{14} cm)

Solar System (Mercury to Neptune): 2.6682×10^{30} grams

As we can see Jupiter carries the majority of the mass, Saturn a pretty large piece, and somewhat large Uranus and Neptune. We don't even list Pluto's mass. When we consider

$$6\int_{\sqrt{2}}^{\sqrt{3}} \frac{1}{t^2}dt = 6\left(\frac{1}{\sqrt{2}} - \frac{1}{\sqrt{3}}\right) = 0.78$$

$$\int_{\sqrt{2}}^{\sqrt{3}} \cos^{-1}(x/2)dx = \frac{1}{6}\left(\sqrt{3}\pi - 6\right) - \frac{\pi - 4}{2\sqrt{2}} = 0.21$$

We are considering $\sqrt{2}$ and $\sqrt{3}$. These come from

$$2\cos(45°) = \sqrt{2}$$

$$2\cos(30°) = \sqrt{3}$$

From 30 degrees to 45 degrees is 15 degrees. The Earth rotates through 360/24 is 15 degrees per hour. The hour is divided into 60 minutes and minutes into 60 seconds…(Next page)

Radius of the Solar System

We have said

$$M = \frac{\pi \rho_0 R^2}{3}$$

For a thin disc. Thus we have a definition for the radius of the solar system, R_s:

Equation 7. $$R_s = \sqrt{\frac{3M_p}{\pi(0.78Ge + 0.21Si)}}$$

Where

Equation 8. $$\frac{1}{\alpha^2 m_p}\sqrt{\frac{h 4\pi r_p^2}{Gc}} \int_{\sqrt{2}}^{\sqrt{3}} \frac{1}{t^2} dt = 0.78$$

Equation 9. $$\int_{\sqrt{2}}^{\sqrt{3}} \cos^{-1}(x/2) dx = 0.21$$

Equation 10. $$air = 0.78 N_2 + 0.21 O_2$$

Equation 11. $$\frac{air}{H_2O} \approx \Phi$$

M_p is the mass of all the planets. We also have that the molar mass of air to the molar mass of water is approximately the golden ratio. The interesting thing is we determine a definition for the radius of the solar system and predict the hydrocarbons (backbones of life) all in one fell swoop.

Proton Planck-Seconds

```
By what value would you like to increment?: 0.01
How many values would you like to calculate for t in equation 1 (no
more than 100?): 100
86.1425 protons 0.070000 seconds 0.142517 decpart
50.2498 protons 0.120000 seconds 0.249805 decpart
43.0713 protons 0.140000 seconds 0.071259 decpart
40.1998 protons 0.150000 seconds 0.199841 decpart
30.1499 protons 0.200000 seconds 0.149876 decpart
26.2173 protons 0.230000 seconds 0.217283 decpart
25.1249 protons 0.240000 seconds 0.124893 decpart
24.1199 protons 0.250000 seconds 0.119900 decpart
23.1922 protons 0.260000 seconds 0.192213 decpart
20.0999 protons 0.300000 seconds 0.099920 decpart
```

```
17.2285 protons 0.350000 seconds 0.228504 decpart
15.0749 protons 0.400000 seconds 0.074944 decpart
14.0232 protons 0.430000 seconds 0.023204 decpart
13.1086 protons 0.460000 seconds 0.108647 decpart
12.0600 protons 0.500000 seconds 0.059957 decpart
```

We see Germanium(32 protons) occurs between 0.15 and 0.20 seconds and silicon (12 protons) occurs at 0.5 seconds. Let's hone the first.

```
By what value would you like to increment?: 0.00355
How many values would you like to calculate for t in equation 1 (no
more than 100?): 100
566.1949 protons 0.010650 seconds 0.194885 decpart
283.0974 protons 0.021300 seconds 0.097443 decpart
113.2390 protons 0.053250 seconds 0.238968 decpart
106.1615 protons 0.056800 seconds 0.161537 decpart
77.2084 protons 0.078100 seconds 0.208389 decpart
53.0808 protons 0.113600 seconds 0.080769 decpart
47.1829 protons 0.127800 seconds 0.182907 decpart
36.1401 protons 0.166850 seconds 0.140114 decpart
32.0488 protons 0.188150 seconds 0.048786 decpart
```

And we see it occurs at 0.188150 seconds. At 47/250.

We see if we take equation 3:

$$\frac{1}{6\alpha^2 m_p}\sqrt{\frac{h 4\pi r_p^2}{Gc}} = 1.004996352 seconds$$

Then where

$$\frac{1}{\alpha^2 m_p}\sqrt{\frac{h 4\pi r_p^2}{Gc}} = (6 protons)(1 second)$$

In Planck-seconds where $5.391247(60)E-44 seconds = 1 Planck Second$

We have:

$$\frac{1}{\alpha^2 m_p}\sqrt{\frac{h 4\pi r_p^2}{Gc}} = 6 proton \cdot seconds$$

$$\frac{1}{\alpha^2 m_p}\sqrt{\frac{h 4\pi r_p^2}{Gc}} = 1E43 proton \cdot planck \cdot seconds$$

Si=9.274291168E42 planck seconds
Ge=3.487133479E42 planck seconds

12Si=1E44 proton planck seconds
32Ge=1E44 proton planck seconds

Where 12 and 32 are the protons in Si and Ge respectively. Concerning equations 7, 8, 9,10, 11, and indeed equation 1, we should be interesting in the ratio 78% and 21% and the way they work with one another.

0.78=39/50
0.21=21/100

Which are basically $\pi/4$ and $(1 - \pi/4)$.

Planets Like An Atom

In so much as we have a definition for the radius of the solar system

$$R_s = \sqrt{\frac{3M_p}{\pi(0.78Ge + 0.21Si)}}$$

$$\frac{1}{\alpha^2 m_p}\sqrt{\frac{h 4\pi r_p^2}{Gc}} \int_{\sqrt{2}}^{\sqrt{3}} \frac{1}{t^2} dt = 0.78$$

$$\int_{\sqrt{2}}^{\sqrt{3}} \cos^{-1}(x/2) dx = 0.21$$

Through life at its fundamental structure the hydrocarbons

$$\frac{1}{t_1}\frac{1}{\alpha^2 m_p}\sqrt{\frac{h 4\pi r_p^2}{Gc}} = 6 protons$$

$$\frac{1}{t_6}\frac{1}{\alpha^2 m_p}\sqrt{\frac{h 4\pi r_p^2}{Gc}} = 1 proton$$

Which occurs in air and water that makes life possible

$$air = 0.78 N_2 + 0.21 O_2$$

$$\frac{air}{H_2O} \approx \Phi$$

We want to consider its ground state, and because Mercury the first planet is so small and carries little energy we proceed to Venus and describe it in terms of silicon and germanium like we did with our definition for the radius of the solar system. I find by molar mass we have its average orbital distance from the sun (0.72 AU) precisely in the following expression:

Equation 12. $$0.72 AU = \frac{1}{Ge^2}\left(\frac{2SiGe + \frac{Si^3}{Ge}}{1 + \frac{Si^2}{Ge^2}}\right)$$

Where Si=28.09 grams/mole and Ge=72.64 grams/mole. If we rewrite this:

Equation 13. $$r(x,y) = \frac{1}{y^2}\left(\frac{2xy + \frac{x^3}{y}}{1 + \frac{x^2}{y^2}}\right)$$

And differentiate

$$\frac{\partial r}{\partial x} = \frac{x^4 + x^2 y^2 + 2y^4}{y(x^2 + y^2)^2}$$

$$\frac{\partial r}{\partial y} = \frac{x(x^4 + x^2 y^2 + 2y^4)}{y^2(x^2 + y^2)^2}$$

And evaluate these for silicon and germanium we have

$$\frac{\partial r}{\partial x} = 0.0022626491$$

$$\frac{\partial r}{\partial y} = 0.008749699$$

Which is:

$$\sqrt{(0.00226)^2 + (0.00875)^2} = 0.009037151 \frac{AU \cdot mol}{gram}$$

Thus doing as we did before, we move down group 14 of the periodic table from carbon, to silicon, to germanium and have:

C+Si+Ge=12.01+28.09+72.64=112.74g/mol

We have

$$0.009037 \frac{AU \cdot mol}{g} \cdot 112.74 \frac{g}{mol} = 1.0 AU$$

Which is Earth Orbit. Plots of our equation for Venus are:

$$r(x, y) = \frac{\frac{x^3}{y} + 2xy}{y^2\left(\frac{x^2}{y^2} + 1\right)}$$

3D plot

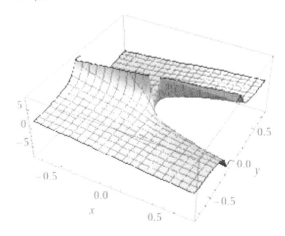

Contour plot

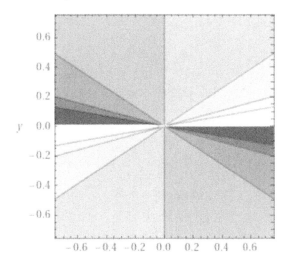

It may be we want to consider Jupiter the ground state as it carries the majority of the mass of the solar system. We may have that Venus is the ground state for the terrestrials planets and Jupiter the ground state for the outer gas giants, as separated by the asteroid belt, the location where a planet cannot form. This takes us to Mars, which is beyond which it lay.

Mars

The asteroid belt is about 1AU thick at 2.2 to 3.2 AU from the sun, the earth-sun separation being 1 AU. Carbon-12 the basis of life as we know it is 12.01 g/mol. It is made in stars from Beryllium-8 is 8.0053051 g/mol. Thus we have

Equation 14. $$\frac{^{12}C}{^{8}Be} = \frac{Mars}{Earth}\Delta x$$

Equation 15. $$\Delta x = 1AU$$

Delta x equal to 1 AU is both Earth and The Asteroids. Mars is at 1.52AU. Delta X cancels with the Earth leaving Mars equal to carbon to beryllium, which is life, Does this say we need to be able to colonize Mars to succeed as a species?

$$^{12}C = {}^{4}He + {}^{8}Be$$

Or, does it mean we need to put bases on the moon to mine Helium-3 as a clean, renewable energy source.

The Moon

Essentially as the moon orbits the earth it makes 12 revolutions for each revolution of the Earth around the sun which is 365.25 days. That is to say

$$T_e = 365.25 days$$

$$T_m = 29.53059 days$$

These are frequencies of

$$f_e = 0.002737851 s^{-1}$$

$$f_m = 0.033863191 days^{-1}$$

In radians per day these are:

$$\omega_e = 0.0172$$

$$\omega_m = 0.21$$

Thus the equations of their phases are:

$$y_e = cos(\omega_e t)$$

$$y_m = (2.57E-3)cos(\omega_m t)$$

Where t is in days and 2.55605E-3= (radius lunar orbit)(radius earth orbit)=384,400km/(149,597,876km).

We can say the frequency of the moon is 0.21/0.0172=12.21 times greater than that of the earth. Thus we have the following plots of lunar phases to earth phases:

Input

$\{y1(t) = \cos(0.0172\,t),\ y2(t) = \cos(0.21\,t)\}$

Plots

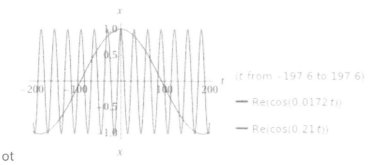

ot

There are 12 moons in a year and 24 hours in a day. Divide twelve by 2 and we have 6, divide 24 by 2 and we have 12. We have:

$$\frac{moons^2}{hours \cdot days} = 6$$

In that days=1, moons-12, hours=24.

At this point we bring-up that remarkable fact the the moon perfectly eclipses the sun. This is because:

$$\frac{(lunar-orbit)}{(earth-orbit)} = \frac{384,400 km}{149,592,870 km} = 0.00257$$

$$\frac{(lunar-radius)}{solar-radius} = \frac{1,738.1}{696,00} = 0.0025$$

Which are approximately equal. As well we can look at it as:

$$\frac{(lunar-radius)}{(lunar-orbit)} = \frac{(1,738.1)}{(384,400)} = 0.00452$$

$$\frac{solar-radius}{earth-orbit} = \frac{696,000}{149,597,870} = 0.00465$$

Which are about the same as well. The interesting thing is that since our ratios are around 0.0025 and 0.0045, then...

$$\frac{0.0045}{0.0025} = \frac{9}{5} = 1.8$$

I say this is interesting because this the ratio of the precious metal gold (Au) to that of silver (Ag) by molar mass:

$$\frac{Au}{Ag} = \frac{196.97}{107.87} = 1.8$$

We have:

$$\frac{(lunar-radius)(earth-orbit)}{(lunar-orbit)^2} = 1.759577590$$

$$\frac{(solar-radius)^2}{(earth-orbit)(lunar-radius)} = 1.863$$

Taking the average between these:

Equation 16. $$\frac{1}{2} \cdot \left(\frac{r_m^2 \cdot R_\odot^2 + r_e^2 R_m^2}{r_m^2 \cdot r_e \cdot R_m} \right) = \frac{Au}{Ag}$$

Where, r_m is the lunar orbit, $R_\odot$ is the solar radius, r_e is the earth orbit, and R_m is the radius of the moon. What this means is that the moon perfectly eclipses the sun because the solar radius is 1.8 times greater than the lunar orbital radius. And interestingly gold is yellow, silver is silver and the sun is gold, and the moon is silver, the moon perfectly eclipses the sun allowing us to study its outer atmosphere, and the exact number of moons in a year is:

$$\frac{365.25}{x} = 29.53059 days$$

x=12.3685304

Is the consecutive integers 0, 1, 2, 3, 4. 5, 6,…8 only missing the 7. Where the equation

$$\frac{(12 moons)^2}{24 hours \cdot 1 day} = 6 \text{ square moons per 24-hour day,…}$$

Is concerned, I believe it is deeply interacting throughout nature with the equation for proton-seconds:

$$\frac{1}{t_1 \alpha^2 m_p} \sqrt{\frac{h 4\pi r_p^2}{Gc}} = 6 protons$$

Because $t_1 = 1 second$

And there are 60 minutes in an hour, and 60 seconds in a minute. We can equate them because while:

$$\frac{1}{t_1}\frac{1}{\alpha^2}\cdot\sqrt{\frac{h4\pi r_p^2}{Gc}}$$

Has units of mass, dividing it by m_p gives a number of protons, but but as well you can think of the m_p cancelling with the mass leaving pure number, and in

$$\frac{(12moons)^2}{24hours\cdot 1day}=6$$

You can think of 12 moons being the number of new moons that appear in the sky per time of the journey of the sun North and back South again marking the seasons. We can divide each day into 24 units and so it may be that there is some dynamic behind:

$$\frac{1}{t_1\alpha^2 m_p}\sqrt{\frac{h4\pi r_p^2}{Gc}}=\frac{(12moons)^2}{24hours\cdot 1day}$$

Meaning:

$$6 protons = \frac{(12moons)^2}{24hours\cdot 1day}=C$$

Theoretical Value For Proton Radius

In that we have

$$\frac{1}{\alpha^2 m_p}\sqrt{\frac{h4\pi r_p^2}{Gc}}\int_{t_1}^{t_2}\frac{1}{t^2}dt=\mathbb{N}$$

And the periodic table of the elements is cyclical with 18 groups and

$$6=\frac{1}{\alpha^2 m_p}\sqrt{\frac{h4\pi r_p^2}{Gc}}$$

Then perhaps we are supposed to write

$$\frac{3}{\alpha^2 m_p}\sqrt{\frac{h4\pi r_p^2}{Gc}}\int_{t_1}^{t_2}\frac{1}{t^2}dt=18$$

In fact, what if the 3 is supposed to be pi, then

$$\frac{\pi}{\alpha^2 m_p}\sqrt{\frac{h 4\pi r_p^2}{Gc}}\int_{t_1}^{t_2}\frac{1}{t^2}dt = 18$$

Then we would say that

k=18/pi=5.7229577951

The parameter in our constant with the most uncertainty is the radius of a proton r_p. If the 3 is supposed to be pi, then the radius of a proton becomes:

$$r_p = k\alpha^2 m_p \sqrt{\frac{Gc}{4\pi h}}$$

Which gives

$$r_p = 8.790587E - 16m$$

About 95% raw most recent value measured. But, if

$$\frac{1}{\alpha^2 m_p}\sqrt{\frac{h 4\pi r_p^2}{Gc}}$$

Is supposed to be 6 and it is supposed to be multiplied by three to give 18 even which we need for chemistry so we have 18 protons in the last group of the periodic table which is important because we need argon with 18 protons for predicting valence numbers of elements in terms of their need to attain noble gas electron configuration. Then we get

$$r_p = 8.288587E - 16m = 0.829fm$$

This is in very close agreement with the most recent value measured which is

$$r_p = 0.833 +/- 0.014$$

Which is good for chemistry.

Part 3: The Wave-Packet

Thus we are interested in:

$$\pi \int_{Si}^{Ge} \int_{Si}^{Ge} (\nabla \times \vec{u}) \cdot d\vec{a} = \prod_{i=1}^{5} x_i$$

$$\prod_{i=1}^{5} x_i = Si \cdot Ge \cdot C \cdot Ag \cdot Au$$

Because we see these elements are a gaussian distribution (See graph on next page) and a wave graph (page 39)…

B. C. N.

Al. Si. P.

Ga. Ge. As.

Ni. Cu. Zn.

Pd. Ag. Cd.

Pt. Au. Hg.

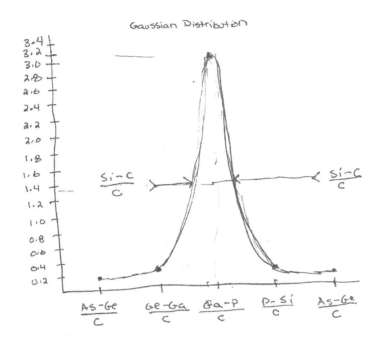

Gaussian Distribution

$\frac{As-Ge}{C} = 0.18996$

$\frac{Ge-Ga}{C} = 0.2428612$

$\frac{Ga-P}{C} = 3.22615$ $\qquad \frac{Si-C}{C} = 1.33827$

$\frac{P-Si}{C} = 0.24051$

$\frac{As-Ge}{C} = 0.18996$

$\frac{As}{C} = \frac{74.9216}{12.011} = 6.23775 \qquad \frac{Ga}{C} = \frac{69.723}{12.011} = 6.04779$

$\frac{Ge}{C} = \frac{72.64}{12.011} = 5.8049288 \qquad \frac{P}{C} = \frac{30.9737620}{12.011} = 2.57878$

$\qquad\qquad\qquad\qquad\qquad\qquad \frac{Si}{C} = \frac{28.085}{12.011} = 2.33827$

We consider a Gaussian wave-packet at t=0:

$$\psi(x,0) = A e^{-\frac{x^2}{2d^2}}$$

We say that d which in quantum mechanics would be the delocalization length when squared is $\frac{Si-C}{C}$. A is the amplitude and we might say it is $\frac{Ga-P}{C}$. We write the wave packet as a Fourier transform which is:

$$\psi(x,0) = A e^{-\frac{x^2}{2d^2}} = \int \frac{dp}{2\pi\hbar} \phi_p e^{\frac{i}{\hbar}px}$$

We use the identity that gives the integral of a quadratic:

$$\int_{-\infty}^{\infty} e^{-\alpha^2 x + \beta x} dx = \sqrt{\frac{\pi}{\alpha}} e^{\frac{\beta^2}{4\alpha}}$$

Solve the equation

$$i\hbar \partial_t \psi(x,t) = \frac{\hat{p}}{2m} \psi(x,t)$$

With the initial condition

$$\psi(x,0) = \int dp \cdot e^{\frac{p^2 d^2}{2\hbar^2}} \cdot e^{-\frac{i}{\hbar}px}$$

A plane wave is the solution:

$$e^{\frac{i}{\hbar}(px - \epsilon(p)t)}$$

Where, $\epsilon(p) = \frac{p^2}{2m}$

The wave-packet evolves with time as

$$\psi(x,t) = \int dp \cdot e^{\frac{p^2 d^2}{2\hbar^2}} \cdot e^{-\frac{i}{\hbar}(px - \frac{p^2}{2m}t)}$$

Calculate the Gaussian integral of dp

$$\alpha = \frac{d^2}{2\hbar^2} + \frac{it}{2m\hbar} \text{ and } \beta = \frac{ix}{\hbar}$$

The solution is:

$$|\psi|^2 = exp\left[-\frac{x^2}{d^2} \cdot \frac{1}{1+t^2/\tau^2}\right] \text{ where } \tau = \frac{md^2}{\hbar}$$

We notice here one of the things you can do with equation 6 of part 2 is integrate from 0.5 sec to 1 sec and you get one which multiplied by the constant which is six yields six. Now look up 0.5 seconds from the data output from the program and it is silicon, then go to one second and it is carbon, thus the integral from silicon in time to carbon in time is carbon. Now consider life as we know it is based on carbon because it has four valence electrons, but it is not based on silicon, which has four valence electrons as well, because in the presence of oxygen it readily forms SiO2 (sand or glass) leaving it unavailable to nitrogen, phosphorus, and hydrogen to make make amino acids the building blocks of life. But silicon can be doped with phosphorus, boron, gallium, and arsenic to make semiconductors -- transistor technology from which we can build artificial life (artificial intelligence, AI). We can integrate over many time ranges to explore millions of more facets to the equation.

$$\frac{1}{\alpha^2 m_p}\sqrt{\frac{h4\pi r_p^2}{Gc}}\int_{t_{Si}}^{t_C}\frac{1}{t^2}dt = 6 = carbon(C)$$

$$t_{Si} = 0.5 seconds$$

$$t_C = 1 second$$

We take our solution for the probability

$$|\psi|^2 = exp\left[-\frac{x^2}{d^2} \cdot \frac{1}{1+t^2/\tau^2}\right]$$

And we say it is:

Equation 1. $$|\psi|^2 = exp\left[-\frac{Cx^2}{(Si-C)} \cdot \frac{1}{1+t^2/\tau^2}\right]$$

Let's say x=1 proton which corresponds to 6 seconds. Then the probability we have a proton is 100% or, the absolute value of psi squared equals 1 means one second meaning we have carbon or, life in other words. But one it does equal:

$$exp\left[\frac{3}{4} \cdot (1proton)^2 \cdot \frac{1}{1+(6seconds)^2}\right] = 1.02$$

We should say

$$d = \frac{Si-C}{C} = \frac{4}{3} \text{ or } d^2 = \frac{16}{9}$$

And is almost exactly one if we average it with the following:

$$exp\left[\frac{3}{4}\cdot(1proton)^2\cdot\frac{1}{(6seconds)^2}\right]=0.979382$$

The way I am using equation 1 is $\tau=d^2$. We have:

Equation 2. $$|\psi|^2=exp\left[-\frac{C^2x^2}{(Si-C)^2}\cdot\frac{1}{1+\left[\frac{\hbar C^2}{m(Si-C)^2}\right]^2 t^2}\right]$$

Thus for hydrogen:

$$|\psi|^2=exp\left[-\frac{9}{16}x^2\cdot\frac{1}{1+\frac{\hbar^2 81}{m^2 256}t^2}\right]$$

$$|\psi|^2=(1)exp\left[-\frac{9}{16}(1proton)^2\cdot\frac{1}{1+\frac{(0.075)81}{(1)256}(6seconds)^2}\right]=74\%.$$

For Carbon:

$$|\psi|^2=(2)exp\left[-\frac{9}{16}(2proton)^2\cdot\frac{1}{1+\frac{(0.075)81}{(4)256}(3seconds)^2}\right]=26\%$$

This is interesting because the Universe is about 74% Hydrogen and 24% Helium, the rest of the elements making up the other 2%. Thus we can say $\hbar^2=0.075$ or $\hbar=0.27386$. We have multiplied the first by 1 for Hydrogen element 1, and the second by 2 for helium element 2. In a sense then, the probabilities represent the probability of finding hydrogen and helium in the Universe. Hydrogen and and much of the helium were made theoretically in the Big Bang of the big bang theory. The other elements were synthesized from these by the stars. Now we move on to the other set of data in:

$$\prod_{i=1}^{5}x_i=Si\cdot Ge\cdot C\cdot Ag\cdot Au$$

Which is an interesting type of wave enveloped by a couple straight lines with different slopes (Next page)…

$$\frac{Au - Ag}{C} = \frac{196.9657 - 107.8682}{12.011} = 7.4179$$

$$\frac{Ag - Ge}{C} = \frac{107.8682 - 72.64}{12.011} = 2.933$$

$$\frac{1}{2}(7.4 - 2.933)(1) = 2.2335$$

$$\frac{Ge - Si}{C} = \frac{72.64 - 28.085}{12.011} = 3.7095$$

$$(1)(2.933) = 2.933$$

$$2.2335 + 2.933 = 5.1665$$

$$\frac{Si - C}{C} = \frac{28.085 - 12.011}{12.011} = 1.338$$

$$\int_{Si-C}^{(Au-Ag)/C} f(n) \, dn = 5.1665$$
$$\frac{Si-C}{C} \quad +6.34754$$
$$= 11.51404$$
$$\uparrow \uparrow \uparrow$$

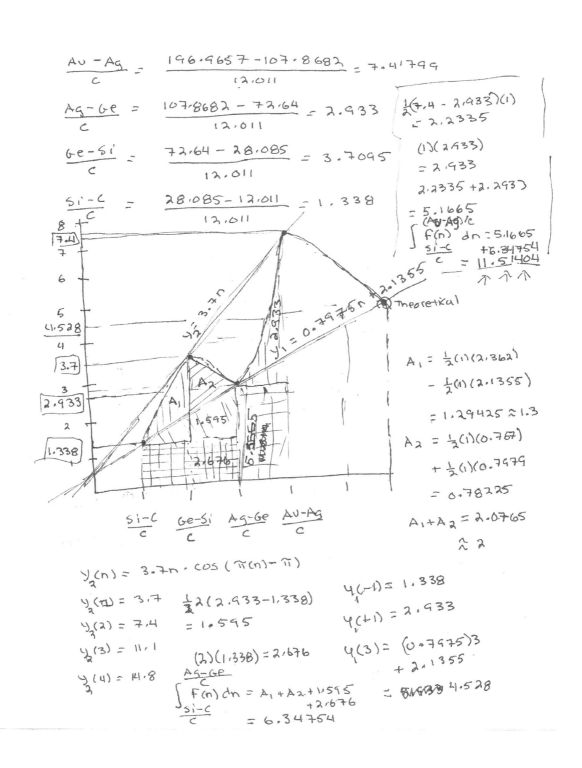

$$A_1 = \frac{1}{2}(1)(2.362) - \frac{1}{2}(1)(2.1355) = 1.29425 \approx 1.3$$

$$A_2 = \frac{1}{2}(1)(0.767) + \frac{1}{2}(1)(0.7979) = 0.78225$$

$$A_1 + A_2 = 2.0765 \approx 2$$

$$y_2(n) = 3.7n \cdot \cos(\pi(n) - \pi)$$

$$y_2(1) = 3.7$$
$$y_2(2) = 7.4$$
$$y_2(3) = 11.1$$
$$y_2(4) = 14.8$$

$$\frac{1}{2}\cdot 2(2.933 - 1.338) = 1.595$$

$$(2)(1.338) = 2.676$$

$$\int_{Si-C}^{Ag-Ge} f(n) \, dn = A_1 + A_2 + 1.595 + 2.676 = 6.34754$$

$$y_1(-1) = 1.338$$
$$y_1(+1) = 2.933$$
$$y_1(3) = (0.7975)3 + 2.1355 = 4.528$$

Thus the graph on page 35 refers to

$$\pi \int_{Si}^{Ge} \int_{Si}^{Ge} (\nabla \times \vec{u}) \cdot d\vec{a}$$

And the graph on page 39 refers to

$$\prod_{i=1}^{5} x_i$$

And the two are equal, which are respectively,…

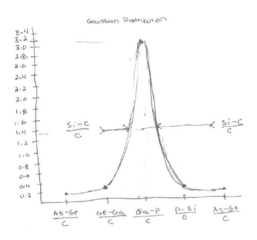

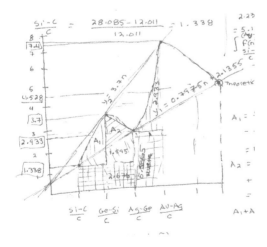

If the first is the probability of finding hydrogen and helium in the Universe determined by a wave function, and the second is a wave, then…

We say the area of the integral as approximated by triangles is:

$$\int_1^3 f(n)dn = A1 + A2 + A3 + R1 + R2 + R3$$

=1.3+0.38+2.933+1.595+2.676+5.1665=12.75

We subtract 3(1.1338) and get 12.75-3.4=9.35

We see that we have come full circle because

$$\left(\frac{Au - Ag}{C} - \frac{Si - C}{C}\right)\Delta n = 18$$

Which are the number of groups in the periodic table.

Indeed we have come full circle because

Equation 3. $\quad 3\left(\dfrac{Au - Ag}{C} - \dfrac{Si - C}{C}\right) = 18$

Equation 4. $\quad \dfrac{3}{\alpha^2 m_p}\sqrt{\dfrac{h4\pi r_p^2}{Gc}}\int_{t_1}^{t_2}\dfrac{1}{t^2}dt = 18$

Where $t_1 = 0.5$ and $t_2 = 1$.

Part 4 Where It Began

Important

	13	14	15
2	B		
3	Al	Si	P
4	Ga	Ge	As

Above we see the artificial intelligence (AI) elements pulled out of the periodic table of the elements. As you see we can make a 3 by 3 matrix of them and an AI periodic table. Silicon and germanium are in group 14 meaning they have 4 valence electrons and want 4 for more to attain noble gas electron configuration. If we dope Si with B from group 13 it gets three of the four electrons and thus has a deficiency becoming positive type silicon and thus conducts. If we dope the Si with P from group 15 it has an extra electron and thus conducts as well. If we join the two types of silicon we have a semiconductor for making diodes and transistors from which we can make logic circuits for AI.

As you can see doping agents As and Ga are on either side of Ge, and doping agent P is to the right of Si but doping agent B is not directly to the left, aluminum Al is. This becomes important. I call (As-Ga) the differential across Ge, and (P-Al) the differential across Si and call Al a dummy in the differential because boron B is actually used to make positive type silicon.

That the AI elements make a three by three matrix they can be organized with the letter E with subscripts that tell what element it is and it properties, I have done this:

$$\begin{pmatrix} E_{13} & E_{14} & E_{15} \\ E_{23} & E_{24} & E_{25} \\ E_{33} & E_{34} & E_{35} \end{pmatrix}$$

Thus E24 is in the second row and has 4 valence electrons making it silicon (Si), E14 is in the first row and has 4 valence electrons making it carbon (C). I believe that the AI elements can be organized in a 3 by 3 matrix makes them pivotal to structure in the Universe because we live in three dimensional space so the mechanics of the realm we experience are described by such a matrix, for example the cross product. Hence this paper where I show AI and biological life are mathematical constructs and described in terms of one another.

We see, if we include the two biological elements in the matrix (E14) and and (E15) which are carbon and nitrogen respectively, there is every reason to proceed with this paper if the idea is to show not only are the AI elements and biological elements mathematical constructs, they are described in terms of one another. We see this because the first row is (B, C, N) and these happen to be the only elements that are not core AI elements in the matrix, except boron (B) which is out of place, and aluminum (Al) as we will see if a dummy representative makes for a mathematical construct, the harmonic mean. Which means we have proved our case because the first row if we take the cross product between the second and third rows are, its respective unit vectors for the components meaning they describe them!

The Computation

$$\vec{A} = (Al, Si, P)$$

$$\vec{B} = (Ga, Ge, As)$$

$$\vec{A} \times \vec{B} = \begin{pmatrix} \hat{B} & \hat{C} & \hat{N} \\ Al & Si & P \\ Ga & Ge & As \end{pmatrix} = (Si \cdot As - P \cdot Ge)\hat{B} + (P \cdot Ga - Al \cdot As)\hat{C} + (Al \cdot Ge - Si \cdot Ga)\hat{N}$$

$$\vec{A} \times \vec{B} = -145\hat{B} + 138\hat{C} + 1.3924\hat{N}$$

$$A = \sqrt{26.98^2 + 28.09^2 + 30.97^2} = 50 g/mol$$

$$B = \sqrt{69.72^2 + 72.64^2 + 74.92^2} = 126 g/mol$$

$$\vec{A} \cdot \vec{B} = AB\cos\theta$$

$$\cos\theta = \frac{6241}{6300} = 0.99$$

$$\theta = 8°$$

$$\vec{A} \times \vec{B} = AB\sin\theta = (50)(126)\sin 8° = 877.79$$

$$\sqrt{877.79} = 29.6 g/mol \approx Si = 28.09 g/mol$$

And silicon (Si) is at the center of our AI periodic table of the elements. We see the biological elements C and N being the unit vectors are multiplied by the AI elements, meaning they describe them! But we have to ask; Why does the first row have boron in it which is not a core biological element, but is a core AI element? The answer is that boron is the one AI element that is out of place, that is, aluminum is in its place. But we see this has a dynamic function.

The Dynamic Function

The primary elements of artificial intelligence (AI) used to make diodes and transistors, silicon (Si) and germanium (Ge) doped with boron (B) and phosphorus (P) or gallium (Ga) and arsenic (As) have an asymmetry due to boron. Silicon and germanium are in group 14 like carbon (C) and as such have 4 valence electrons. Thus to have positive type silicon and germanium, they need doping agents from group 13 (three valence electrons) like boron and gallium, and to have negative type silicon and germanium they need doping agents from group 15 like phosphorus and arsenic. But where gallium and arsenic are in the same period as germanium, boron is in a different period than silicon (period 2) while phosphorus is not (period 3). Thus aluminum (Al) is in boron's place. This results in an interesting equation.

Equation 1. $$\frac{Si(As - Ga) + Ge(P - Al)}{SiGe} = \frac{2B}{Ge + Si}$$

The differential across germanium crossed with silicon plus the differential across silicon crossed with germanium normalized by the product between silicon and germanium is equal to the boron divided by the average between the germanium and the silicon. The equation has nearly 100% accuracy (note: using an older value for Ge here, it is now 72.64 but that makes the equation have a higher accuracy):

$$\frac{28.09(74.92 - 69.72) + 72.61(30.97 - 26.98)}{(28.09)(72.61)} = \frac{2(10.81)}{(72.61 + 28.09)}$$

$$0.213658912 = 0.21469712$$

$$\frac{0.213658912}{0.21469712} = 0.995$$

Due to an asymmetry in the periodic table of the elements due to boron we have the harmonic mean between the semiconductor elements (by molar mass):

Equation 2. $\quad \frac{Si}{B}(As - Ga) + \frac{Ge}{B}(P - Al) = \frac{2SiGe}{Si + Ge}$

This is Stokes Theorem if we approximate the harmonic mean with the arithmetic mean:

$$\int_S (\nabla \times \vec{u}) \cdot d\vec{S} = \oint_C \vec{u} \cdot d\vec{r}$$

$$\int_0^1 \int_0^1 \left[\frac{Si}{B}(As - Ga) + \frac{Ge}{B}(P - Al) \right] dxdy \approx \frac{1}{Ge - Si} \int_{Si}^{Ge} xdx$$

We can make this into two integrals:

Equation 3. $\quad \int_0^1 \int_0^1 \frac{Si}{B}(As - Ga) dydz \approx \frac{1}{3} \frac{1}{(Ge - Si)} \int_{Si}^{Ge} xdx$

Equation 4. $\quad \int_0^1 \int_0^1 \frac{Ge}{B}(P - Al) dxdz \approx \frac{2}{3} \frac{1}{(Ge - Si)} \int_{Si}^{Ge} ydy$

If in the equation (The accurate harmonic mean form):

Equation 5. $\quad \frac{Si}{B}(As - Ga) + \frac{Ge}{B}(P - Al) = \frac{Ge - Si}{\int_{Si}^{Ge} \frac{dx}{x}}$

We make the approximation

Equation 6. $\quad \frac{2SiGe}{Si + Ge} \approx Ge - Si$

Then the Stokes form of the equation becomes

Equation 7. $$\int_0^1 \int_0^1 \left[\frac{Si}{B}(As - Ga) + \frac{Ge}{B}(P - Al) \right] dydz = \int_{Si}^{Ge} dx$$

Thus we see for this approximation there are two integrals as well:

Equation 8. $$\int_0^1 \int_0^1 \frac{Si}{B}(As - Ga) dydz = \frac{1}{3} \int_{Si}^{Ge} dz$$

Equation 9. $$\int_0^1 \int_0^1 \frac{Ge}{B}(P - Al) dydz = \frac{2}{3} \int_{Si}^{Ge} dz$$

For which the respective paths are

$$y_1 = \frac{1}{3} \frac{B}{SiGa} ln(z)$$

$$y_2 = \frac{2}{3} \frac{B}{SiAl} ln(z)$$

One of the double integrals on the left is evaluated in moles per grams, the other grams per mole (0 to 1 moles per gram and 0 to 1 grams per mole).

By making the approximation

$$\frac{2SiGe}{Si + Ge} \approx Ge - Si$$

In

$$\frac{Si(As - Ga)}{B} + \frac{Ge(P - Al)}{B} = \frac{2SiGe}{Si + Ge}$$

We have

Equation 10. $$Si \frac{\Delta Ge}{\Delta S} + Ge \frac{\Delta Si}{\Delta S} = B$$

$\Delta Si = P - Al$ is the differential across Si, $\Delta Ge = As - Ga$ is the differential across Ge and $\Delta S = Ge - Si$ is the vertical differential.

Which is Ampere's Circuit Law

$$\nabla \times \vec{B} = \mu_0 \vec{J} + \mu_0 \epsilon \frac{\partial \vec{E}}{\partial t}$$

We see if written

$$Si \frac{\Delta Ge}{\Delta S} = B - Ge \frac{\Delta Si}{\Delta S}$$

Which is interesting because it is semiconductor elements by molar mass which are used to make circuits.

We say Φ (Phi) is given by

$$a = b + c \quad \text{and} \quad \frac{a}{b} = \frac{b}{c}$$

And

$$\Phi = a/b = 1.618$$

$$\phi = b/a = 0.618$$

ϕ (phi) the golden ratio conjugate. We also find

Equation 11. $\quad (\phi)\Delta Ge + (\Phi)\Delta Si = B$

Thus since

$$\nabla \times \vec{B} = \mu \vec{J} + \mu \epsilon_0 \frac{\partial \vec{E}}{\partial t}$$

$$Si \frac{\Delta Ge}{\Delta S} = B - Ge \frac{\Delta Si}{\Delta S}$$

And we have

Equation 12. $\quad \Delta Ge = \frac{\Delta S}{Si} B - \frac{Ge}{Si} \Delta Si$

$$\left(\nabla^2 - \frac{1}{c^2} \frac{\partial^2}{\partial t} \right) \vec{E} = 0$$

$$\left(\nabla^2 - \frac{1}{c^2} \frac{\partial^2}{\partial t} \right) \vec{B} = 0$$

$$c = \frac{1}{\sqrt{\epsilon_0 \mu}} \approx \phi$$

We see μ and ϵ_0 are both Φ and c is ϕ in the Si (silicon) field wave, but for E and B fields c is the speed of light.

$$\epsilon_0 = 8.854E - 12 F \cdot m^{-1}$$

$$\mu = 1.256E - 6 H/m$$

$$\frac{Ge}{Si} = \mu \epsilon_0$$

$$\frac{\Delta S}{Si} = \mu$$

$$\left(\nabla^2 - \frac{1}{\phi^2} \frac{\partial^2}{\partial x} \right) \overrightarrow{Si} = 0$$

$$\left(\nabla^2 - \frac{1}{\phi^2} \frac{\partial^2}{\partial x} \right) \overrightarrow{Ge} = 0$$

To find the Si wave our differentials are

$$\Delta C = N - B = 14.01 - 10.81 = 3.2$$

$$\Delta Si = P - Al = 30.97 - 26.98 = 3.99$$

$$\Delta Ge = As - Ga = 74.92 - 69.72 = 5.2$$

$$\Delta Sn = Bi - In = 121.75 - 114.82 = 6.93$$

$$\Delta Pb = Bi - Tl = 208.98 - 204.38 = 4.6$$

It is amazing how accurately we can fit these differentials with and exponential equation for the upward increase. The equation is

$$y(x) = e^{0.4x} + 1.7$$

$$y(x) = e^{\frac{2}{5}x} + \frac{17}{10}$$

This is the halfwave:

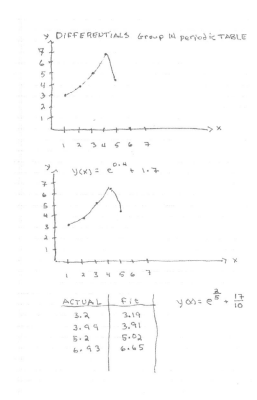

$$y(x) = e^{0.4x} + 1.7$$

$$y(x) = e^{\frac{2}{5}x} + \frac{17}{10}$$

Equation 13. $\quad y(x) = e^{\frac{B}{Al}x} + \frac{Ag}{Cu}$

$$\frac{B}{Al} = \frac{10.81}{26.98} = 0.400667$$

$$\frac{Ag}{Cu} = \frac{107.87}{63.55} = 1.6974 \approx 1.7$$

Interestingly, the 0.4 is boron (B) over aluminum (Al) the very two elements that lead us to looking for a wave equation because boron was the out of place element in the AI periodic table that lead to us using aluminum as its dummy representative in the Si differential and that itself divided into the left hand terms to give us the harmonic mean between the central AI elements semiconductor materials Si and Ge. The Ag and Cu are the central malleable, ductile, and conductive metals used in making electrical wires to carry a current in AI circuitry.

Part 5: Archaeology of other Star Systems

Abstract:

A lot comes together for the Earth orbiting the sun to let us know there is a mystery before us if we look at the archaeology of Earth's astronomy. Thus, does a lot come together considering the archaeology of other star systems as well that can indicate to its intelligent life that they are part of something larger than themselves, as well.

Star System Archaeology

Kepler's Law of planetary motion is

Equation 1. $\qquad T^2 = a^3$

For the Sun with T the orbital period of the planet and a its distance from the Sun in astronomical units, for circular orbits. For other stars we have to include a constant k involving the masses of the bodies:

Equation 2. $\qquad k = G\dfrac{M+m}{4\pi^2} \approx G\dfrac{M}{4\pi^2}$

If the mass of the body orbiting the Star m is is small compared to the mass of the star it is orbiting we have

Equation 3. $\qquad a^3 = \dfrac{GM}{4\pi^2}T^2$

The Earth is the third planet, is in the habitable zone, and its distance from the Sun defines 1 AU. Thus we ask: What is k for other star systems? For stars on the main sequence their luminosity is proportional to their masses raised to the power of 3.5 as an estimate. We have:

Equation 4: $\qquad L = \left(\dfrac{M}{M_\odot}\right)^{3.5}$

Where L is in solar luminosities and M is in solar masses:

$L_\odot = 3.9E26 J/s$

$M_\odot = 1.98847E30 kg$

Further if we say since the Earth is in the right zone to be habitable (H) then if a star is 100 times brighter than the Sun by the inverse square law its habitable zone is $\sqrt{100} = 10 AU$ is 10 times further from the star it orbits than the Earth is from the Sun. We have

Equation 5: $\qquad H = \sqrt{\left(\dfrac{L}{L_\odot}\right)}$

Combining equations 4 and 6:

Equation 6: $\qquad H = \sqrt{\left(\dfrac{M}{M_\odot}\right)^{3.5}}$

For another star system we can write equation 3

Equation 7. $$\frac{T^2}{a^3} = nk$$

Where n is the number of solar masses of the star. Combining this with equation 6:

Equation 8. $$T^2 = nk\left(\frac{M}{M_\odot}\right)^{\frac{21}{4}}$$

Thus if the star is 100 times more luminous than the sun

$ln(100) = 3.5 ln(M/M_\odot)$

Or,…

$$\frac{M}{M_\odot} = 3.72759$$

Yielding from equation 6

$\mathbb{H} = \sqrt{3.72759^{3.5}} = 10 AU$

And

Equation 9. $$T = \left(\frac{M}{M_\odot}\right)^{5/2}$$

Which is 61 years for 3.72759. This result is (61 yr)(365.25 days)=22,280.25 days for the orbital period. 10 AU from this star is about the distance Saturn is from the Sun (9.54AU), with year equal to 10,759 days. The orbital period is longer because it orbits a bit further from its star. If in the planet that has life orbiting a star has an indication to its intelligence that there is a mystery before it, just as we do on the Earth in that we have a moon that perfectly eclipses the Sun as seen from the Earth, then…

This is because:

$$\frac{(lunar-orbit)}{(earth-orbit)} = \frac{384,400 km}{149,592,870 km} = 0.00257$$

$$\frac{(lunar-radius)}{solar-radius} = \frac{1,738.1}{696,00} = 0.0025$$

Which are approximately equal. As well we can look at it as:

$$\frac{(lunar-radius)}{(lunar-orbit)} = \frac{(1,738.1)}{(384,400)} = 0.00452$$

$$\frac{solar-radius}{earth-orbit} = \frac{696,000}{149,597,870} = 0.00465$$

Which are about the same as well. The interesting thing is that since our ratios are around 0.0025 and 0.0045, then…

$$\frac{0.0045}{0.0025} = \frac{9}{5} = 1.8$$

I say this is interesting because this the ratio of the precious metal gold (Au) to that of silver (Ag) by molar mass these elements being used for religious and ceremonial jewelry:

$$\frac{Au}{Ag} = \frac{196.97}{107.87} = 1.8$$

We have:

$$\frac{(lunar-radius)(earth-orbit)}{(lunar-orbit)^2} = 1.759577590$$

$$\frac{(solar-radius)^2}{(earth-orbit)(lunar-radius)} = 1.863$$

Taking the average between these:

Equation 10.
$$\frac{1}{2} \cdot \left(\frac{r_m^2 \cdot R_\odot^2 + r_e^2 R_m^2}{r_m^2 \cdot r_e \cdot R_m} \right) = \frac{Au}{Ag}$$

Where, r_m is the lunar orbit, $R_\odot$ is the solar radius, r_e is the earth orbit, and R_m is the radius of the moon. What this means is that the moon perfectly eclipses the sun because the solar radius is 1.8 times greater than the lunar orbital radius. And interestingly gold is yellow, silver is silver and the sun is gold, and the moon is silver, the moon perfectly eclipses the sun allowing us to study its outer atmosphere.

The Calendar of a People's Star System

By dividing the day into 24 hours, the hour into 60 minutes, and the minute into 60 seconds, the second is 1/86400 of day. By doing this we have a twelve-hour daytime at spring and fall equinox on the equator, 12 being the most divisible number for its size (smallest abundant number). That is to say that twelve is evenly divisible by 1,2,3,4,6 which precede it and 1+2+3+4+6=16 is greater than twelve. As such there is about one moon per 30 days and about 12 moons per year (per each orbit) giving us a twelve-month calendar. This is all further convenient in that the moon and earth are in very close to circular orbits and the circle is evenly divisible by 30, 45, 60, and 120 if we divide the circle into 360 degrees which are special angles very useful to the workings of physics and geometry. Further, the 360 degrees of a circle are about the 365 days of a year (period of one earth orbit) so as such the earth moves through about a degree a day in its journey around the sun. Thus, through these observations down through the ages since ancient times we have constructed the duration of a second wisely enough to make a lot work together.

Essentially as the moon orbits the earth it makes 12 revolutions for each revolution of the Earth around the sun which is 365.25 days. That is to say

$T_e = 365.25 days$

$T_m = 29.53059 days$

These are frequencies of

$f_e = 0.002737851 s^{-1}$

$f_m = 0.033863191 days^{-1}$

In radians per day these are:

$\omega_e = 0.0172$

$\omega_m = 0.21$

Thus the equations of their phases are:

$y_e = cos(\omega_e t)$

$y_m = (2.57E-3)cos(\omega_m t)$

Where t is in days and 2.55605E-3=
(radius lunar orbit)(radius earth orbit)=384,400km/(149,597,876km).

We can say the frequency of the moon is 0.21/0.0172=12.21 times greater than that of the earth. Thus we have the following plots of lunar phases to earth phases:

Input

$\{y1(t) = \cos(0.0172\,t),\ y2(t) = \cos(0.21\,t)\}$

Plots

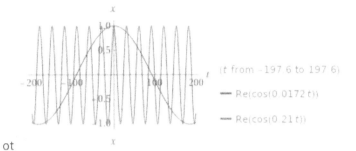

ot

There are 12 moons in a year and 24 hours in a day. Divide twelve by 2 and we have 6, divide 24 by 2 and we have 12. We have:

$$\frac{moons^2}{hours \cdot days} = 6$$

In that days=1, moons-12, hours=24.

The Geometrical Explanation of Seconds

We have suggested the second is important as well in terms of the phases of the moon and the earth and that these determined the calendar system we use. We further suggested there is a connection of this to the structure found in geometry, and this is what we want to explore further, here. We have all of this can be compactly written as:

$$2cos(\pi/n) = 1, \sqrt{2}, \Phi, \sqrt{3}, \ldots$$

Where n=3, 4, 5, ,6 ,...

We could evaluate this for n equal to other integers, or even the numbers, but these produce the special triangles, and geometries we are most interested. Thus we will begin by pointing out that

Equation 11.   3 X 4 X 5 X 6 = 360

The amount of degrees into which we divide a circle and that, as such it approximates the number of days in a year (1 revolution of the earth around the sun) and thus we see that

Equation 12. $2cos(\pi/n) = 1, \sqrt{2}, \Phi, \sqrt{3}, \ldots$

Represents days as well (The earth moves through about 1 degree a day in its orbit around the sun) by solving it for n this and the following are what might be characteristic equations of habitable star systems:

Equation 13. $$days = \cos^{-1}(y/2)$$

Where,

$$y = 1, \sqrt{2}, \frac{\sqrt{5}+1}{2}, \sqrt{3}, \ldots$$

Which correspond respectively to:

$n = 3, 4, 5, 6, \ldots$

Which are the unit triangle, unit the square, the unit regular pentagon, and the unit regular hexagon.

The Arrival of Mammalian Life

The asteroid belt seems to have been the source of an asteroid that hit Earth in the Yucatan 65 million years ago leading to the extinction of the dinosaurs and giving mammalian life a chance to evolve into intelligence. Is this part of a natural process that takes place in other star systems?

The asteroid belt is about 1AU thick at 2.2 to 3.2 AU from the sun, the earth-sun separation being 1 AU. Carbon-12 the basis of life as we know it is 12.01 g/mol. It is made in stars from Beryllium-8 is 8.0053051 g/mol. Thus we have

Equation 14. $$\frac{^{12}C}{^{8}Be} = \frac{Mars}{Earth}\Delta x$$

Equation 15. $$\Delta x = 1AU$$

Delta x equal to 1 AU is both Earth and The Asteroids. Mars is at 1.52AU. Delta X cancels with the Earth leaving Mars equal to carbon to beryllium, which is life, Does this say we Mars was made available to Humans to colonize?

$$^{12}C = {}^{4}He + {}^{8}Be$$

Or, does it mean we need to put bases on the moon to mine Helium-3 as a clean, renewable energy source.

Application of the Theory

Kepler-442b is a larger than earth rocky planet orbiting a K-type main sequence star with a mass of 0.61 solar masses, luminosity 12% that of the sun and 0.60 solar radius. It is about 2.9 billion years old compared to thus Sun at 4.6 billion years. It is at a distance of about 1,206 light years in the constellation Lyra. The planet orbits at 0.409 AU with an orbital period of 112.3 days (112 days 9 hours 10 minutes 24 seconds). It is 2.3 earth masses and 1.34 earth radius. It receives 70% the light the earth receives from the sun.

As it would turn-out this exoplanet is tidally locked with the star it orbits being closer in to its star than the earth making its day perhaps weeks to months long. In so far as we don't know the length of its day we cannot determine the orbital period of any moon it might have in terms of such a day. Nor do we even know if it has a moon to define a month from which we can infer the number of months in a year on the planet. All we really know is it has a year 112.3 earth days.

However, let us speculate upon a scenario from what we have developed so far. Equation 9 was derived with the condition:

Equation 16. $$\frac{R_\odot}{r_m} = \frac{Au}{Ag}$$

We might formulate another scenario, While gold (Au) and silver (Ag) are metal used for making religious and ceremonial jewelry so are silver (Ag) and copper (Cu). The thing all of these have in common is that they are the finest malleable and ductile metals used fro the purpose of making electrical wire as they are the finest conductors. Copper is among one of the first metals used for the purpose of adornments and tools because it is soft enough to pound-out without heating it, which was the method first discovered by our ancient ancestors. It was used before smelting (separating from its ores) in its naturally occurring form. We thus have:

Equation 17. $$\frac{R_\odot}{r_m} = \frac{Ag}{Cu}$$

Putting in the radius of the star as 0.6 the radius of the sun which is 696,000km and the molar mass of silver is 107.87 and copper is 63.55 we have $r_m = 246,023.3 km$ as the orbital distance for the moon of Kepler-44b. Thus with the astronomical unit of the the moon (LAU) being one corresponds to a period of 29.3 days. We have

$$\frac{R_\odot}{r_m} = 1.6974$$

$$T = \sqrt{\left(\frac{246,023.3}{384,399}\right)^3} = 0.512$$

$$T = (0.512)(29.53) = 15.1209 days$$

The period of the moon would be almost exactly 15 days. This connects the moon of this planet to the moon of our planet in that the moon goes through about 15 degrees in an hour (360/24=15). This gives credence to something I had considered earlier…

Ethnomusicology is a sub-discipline in archaeology. It studies the history of humans through its musical instruments and music theory. When we discussed the 12 month year based on twelve moons per earth orbital period, we noticed it coincided with elegant mathematics because 12 was evenly divisible by 1, 2, 3, 4, 6 making it abundant. 12 in flamenco is a form of six. 6/8 is a form common to may cultures and a favorite for being lively and dynamic throughout India, the Middle East, and Africa. But perhaps the most beloved in India is Tin Tal, a cycle of 16 which is four, four times. Four is often play four times throughout the Earth's cultures. Thus, if a planet orbited a star that had sixteen moons per its year, the inhabitants might divide the circle into

400 degrees and have four four month seasons rather than four three month seasons. This results in the following table:

12 month year

$$\frac{360}{30} = 12 \qquad 2\cos(30) = \sqrt{3}$$

$$\frac{360}{36} = 10 \qquad 2\cos(36) = \phi$$

$$\frac{360}{45} = 8 \qquad 2\cos(45) = \sqrt{2}$$

$$\frac{360}{60} = 6 \qquad 2\cos(60) = 1$$

$$\frac{360}{90} = 4 \qquad 2\cos(0) = 0$$

16 month year

$$\frac{400}{x} = 12 \qquad x = 100/3 \qquad \sqrt{3}$$

$$\frac{400}{x} = 10 \qquad x = 40 \qquad \phi$$

$$\frac{400}{x} = 8 \qquad x = 50 \qquad \sqrt{2}$$

$$\frac{400}{x} = 6 \qquad x = \frac{200}{3} \qquad 1$$

$$\frac{40}{x} = 4 \qquad x = 100 \qquad 0$$

The luminosity of the sun is:

$$L_0 = 3.9 \times 10^{26} J/s$$

The separation between the earth and the sun is:

$$1.5 \times 10^{11} m$$

The solar luminosity at the earth is reduced by the inverse square law, so the solar constant is:

$$S_0 = \frac{39 \times 10^2}{4\pi(1.5 \times 10^{11})} = 1,370 watts/meter^2$$

That is the effective energy hitting the earth per second per square meter. This radiation is equal to the temperature, T_e to the fourth power by the steffan-bolzmann constant, sigma (σ), T_e can be called the temperature entering, the temperature entering the earth.

S_0 intercepts the earth disc, πr^2, and distributes itself over the entire earth surface, $4\pi r^2$, while 30% is reflected back into space due to the earth's albedo, a, which is equal to 0.3, so

$$\sigma T_e^4 = \frac{S_0}{4}(1-a)$$

$$(1-a)S_0 \left(\frac{\pi r^2}{4\pi r^2}\right)$$

But, just as the same amount of radiation that enters the system, leaves it, to have radiative equilibrium, the atmosphere radiates back to the surface so that the radiation from the atmosphere, σT_a^4 plus the radiation entering the earth, σT_e^4 is the radiation at the surface of the earth, σT_s^4. However,

$$\sigma T_a^4 = \sigma T_e^4$$

And we have…

$$\sigma T_s^4 = \sigma T_a^4 + \sigma T_e^4 = 2\sigma T_e^4$$

$$T_s = 2^{\frac{1}{4}} T_e$$

$$\sigma T_e^4 = \frac{S_0}{4}(1-a)$$

$$\sigma = 5.67 \times 10^{-8}$$

$$S_0 = 1,370$$

a=0.3

$$\frac{1{,}370}{4}(0.7) = 239.75$$

$$T_e^4 = \frac{239.75}{5.67 \times 10^{-8}} = 4.228 \times 10^9$$

$$T_e = 255 Kelvin$$

So, for the temperature at the surface of the Earth:

$$T_s = 2^{\frac{1}{4}} T_e = 1.189(255) = 303 Kelvin$$

Let's convert that to degrees centigrade:

Degrees Centigrade = 303 - 273 = 30 degrees centigrade

And, let's convert that to Fahrenheit:

Degrees Fahrenheit = 30(9/5)+32=86 Degrees Fahrenheit

In reality this is warmer than the average annual temperature at the surface of the earth, but in this model, we only considered radiative heat transfer and not convective heat transfer. In other words, there is cooling due to vaporization of water (the formation of clouds) and due to the condensation of water vapor into rain droplets (precipitation or the formation of rain).

The program in C is as follows:

```c
#include <stdio.h>
#include<math.h>

int main(int argc, const char * argv[]) {
    
    {
        float s, a, l, b, r, AU, N, root, number, answer, C, F;
        printf("We determine the surface temperature of a planet.\n");
        printf("What is the luminosity of the star in solar luminosities? ");
        scanf("%f", &s);
        printf("What is the albedo of the planet (0-1)?" );
        scanf("%f", &a);
        printf("What is the distance from the star in AU? ");
        scanf("%f", &AU);
        r=1.5E11*AU;
        l=3.9E26*s;
        b=l/(4*3.141*r*r);
        
        N=(1-a)*b/(4*(5.67E-8));
```

```
    root=sqrt(N);
    number=sqrt(root);
    answer=1.189*(number);
    printf("\n");
    printf("\n");
    printf("The surface temperature of the planet is: %f K\n",
answer);
    C=answer-273;
    F=(C*1.8)+32;
    printf("That is %f C, or %f F", C, F);
    printf("\n");
    float joules;
    joules=(3.9E26*s);
    printf("The luminosity of the star in joules per second is:
%.2fE25\n", joules/1E25);
    float HZ;
    HZ=sqrt(joules/3.9E26);
    printf("The habitable zone of the star in AU is: %f\n", HZ);
    printf("Flux at planet is %.2f times that at earth.\n", b/1370);
    printf("That is %.2f Watts per square meter\n", b);
    printf("\n");
    printf("\n");
    }
    return 0;
}
```

Let us run our program for Kepler-442b

```
Here we use a single atmospheric layer with no
convection for the planet to be in an equilibrium
state.  That is to say, the temperature stays
steady by heat gain and loss with radiative
heat transfer alone.
The habitable zone is calculated using the idea
that the earth is in the habitable zone for a
star like the Sun. That is, if a star is 100
times brighter than the Sun, then the habitable
zone for that star is ten times further from
it than the Earth is from the Sun because ten
squared is 100.

We determine the surface temperature of a planet.
What is the luminosity of the star in solar luminosities? 0.12
What is the albedo of the planet (0-1)?0.3
What is the distance from the star in AU? 0.409


The surface temperature of the planet is: 279.523163 K
That is 6.523163 C, or 43.741692 F
```

```
The luminosity of the star in joules per second is: 4.68E25
The habitable zone of the star in AU is: 0.346410
Flux at planet is 0.72 times that at earth.
That is 989.67 Watts per square meter
```

Ideally if the nearest star to us was similar to our Sun it would be best to start there. And indeed the nearest star system is composed of a three stars and two of them are comparable to the sun, in fact almost identical. But being a triple system, with which one of the stars has been found to have a planet in its habitable zone, it can only be solved with equations other than ours which work only for a single star with planets small in comparison to to its mass. The triple star system I speak of is Alpha Centauri A, B, and C.

Alpha Centauri is the nearest of the stars to us and is a triple system designated Alpha Centauri A, B, And C. Alpha Centauri A and B are much like the sun, similar in mass and luminosity. Alpha Centauri A is is 1.1 solar masses and Alpha Centauri B is 0.907 solar masses. They orbit a common center, which can be taken as the focus of an ellipse at the center of gravity between the two. The period is 79.91 years. The eccentricity of the orbit is high and is given as e=0.519.

Alpha Centauri A is thought to have a Neptune sized planet in its habitable zone. While A and B orbit one another their closest approach is 11.2 AU, and furtherest is 35.6 AU, but C (Proxima Centauri) has its orbit very far from these two at 0.21 light years, a whopping 13,000 AU. The 11.2 and 35.6 AU for A and B are distances like Saturn and Pluto from the Sun, respectively.

With the launching of the James Webb Space Telescope a lot of new data should be coming in for planets and their moons, and we will have more to work with.

The Author

Made in the USA
Las Vegas, NV
01 February 2022

42773668R00037